How America Can Lose the Fourth Industrial Revolution

David P. Goldman
Washington Fellow
The Claremont Institute's CENTER FOR THE AMERICAN WAY OF LIFE

PROVOCATIONS #2
CLAREMONT INSTITUTE
CENTER FOR THE AMERICAN WAY OF LIFE

All rights reserved. No part of this publication may be re-produced or transmitted in any form or by any means electronic or mechanical, including photocopy, recording, or any in-formation storage and retrieval system now known or to be in-vented, without permission in writing from the publisher, except by a reviewer who wishes to quote brief passages in connection with a review written for inclusion in a magazine, newspaper, website, or broadcast.

Design: David Reaboi/Strategic Improvisation

ISBN: 9798494738264

Published on in the United States by the Claremont Institute, November 11, 2021.

Manufactured in the United States of America

WHAT IF CHINA WINS THE RACE TO THE FOURTH INDUSTRIAL REVOLUTION?

We are currently in the midst of a Fourth Industrial Revolution, defined by metadata and artificial intelligence systems and applications. The Third Industrial Revolution, based on computation and communications, was driven by the United States. China wants to lead the Fourth Industrial Revolution and thereby win the future. It may succeed in doing so. If it does, the consequences for the United States will be disastrous: we will become considerably poorer, our politics will be less stable, and our economy will be dominated by an adversary. Moreover, the military dimension of this revolution could make obsolete many, if not all, of our core weapons systems.

Refusing to see these obvious possibilities, the neoliberal economic consensus that dominates elite Washington thinking, as well as most of Silicon Valley and Wall Street, still hopes or pretends to hope that a leap in China's productivity would only benefit the United

States by providing cheaper and more varied goods and services in trade.

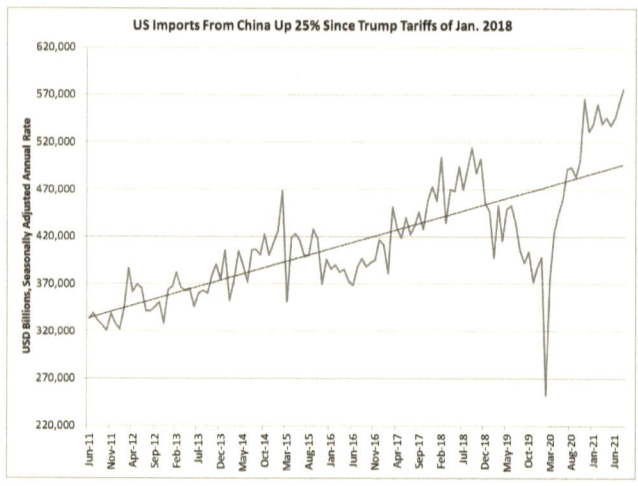

Yet we live in a winner-takes-all world. America's wealth, as well as its financial stability, depends extensively on technological leadership, which has created most of the new wealth in the United States during the past two decades. Chinese leadership in the Fourth Industrial Revolution would precipitate the unraveling of America's global financial position and create a profound and systemic crisis.

The United States now imports almost $600 billion a year of Chinese goods, 25 percent more than in January of 2018 when President Trump imposed punitive tariffs. That is equal to about a quarter of US manufacturing GDP. Far from decoupling from China, a widespread proposal during the COVID-19 pandemic, the United States has coupled itself to China more closely than ever.

That is the result of more than $5 trillion in fiscal stimulus—a boost to demand more than triple the fiscal support for the economy during the 2008-9 Great Recession—without corresponding investment in US production capacity. The United States is now running a trillion-dollar current account deficit, while the federal budget deficit stands at more than 10 percent of GDP, according to the Congressional Budget Office. Chinese dominance in the next generation of manufacturing and logistics will erode America's position as the provider of the world's dominant reserve currency and ultimately lead to a funding crisis for America's burgeoning internal and external debt.

WHAT IS THE FOURTH INDUSTRIAL REVOLUTION?

As World Economic Forum founder Klaus Schwab wrote in 2016:

> The First Industrial Revolution used water and steam power to mechanize production. The Second used electric power to create mass production. The Third used electronics and information technology to automate production. Now a Fourth Industrial Revolution is building on the Third, the digital revolution that has been occurring since the middle of the last century. It is characterized by a fusion of technologies that is blurring the lines between the physical, digital, and biological spheres.[1]

There is now an extensive body of literature on the Fourth Industrial Revolution and its definitive technology, artificial intelligence.[2] Technologies already in place or in advanced stages of development include:

- networks of industrial robots that communicate by mobile broadband and program themselves for manufacturing tasks;

- public health systems that upload in real time the vital signs of hundreds of millions of individuals and correlate this data with detailed information on the location of COVID-19 cases, using artificial intelligence to predict prospective future outbreaks, and to direct forensic testing to potential hotspots;

- AI-based diagnosis of diseases, telemedicine, and surgery by remote-controlled robots using high-speed, low-latency broadband;

- "smart logistics," including fully automated ports where automated cranes find individual containers and move them to autonomous trucks;

- AI-driven pharmaceuticals research using datasets that include the digitized health records, sequenced DNA, and real-time vital signs of hundreds of millions of subjects;

- "smart farms," where sensors at the base of plants send data on moisture, pesticide, and fertilizer needs over broadband to drones that deliver the required inputs; and

- digital finance that employs big datasets to evaluate household and microenterprise credit to make loans and transmit payments.

American leadership of the Third Industrial Revolution in computation and communications made this country a magnet for the world's capital.[3] The United States was able to run a chronic deficit in goods and services because a world hungry for investments wanted our assets. Substituting cheap imports for domestic manufacturing may have been a policy error, but it was an affordable policy error because of America's leadership in the digital age. To put this in perspective: the combined market capitalization of America's mega cap technology stocks (Facebook, Apple, Amazon, Microsoft, Netflix, and Google) rose to $9 trillion in 2021 from $1 trillion in 2013—that is, to 26 percent of the market capitalization of the S&P 500 from 8 percent in 2013.[4]

In the digital world of the Third Industrial Revolution network effects fostered giant "natural" monopolies. There is only one Google, not Google and Altavista; one Microsoft Word, not MS Word and WordPerfect; one Amazon, not Amazon and Walmart. In the Fourth Industrial Revolution, where natural monopolies will arise from dominance in big data, China has a double advantage. First, it has put massive state resources into frontier R&D, and it is producing seven times as many STEM bachelor's degrees as the United States. Second, China's political system allows the harvesting of personal data, including medical records, without the constraint of Western privacy laws, which gives it enormous advantages. Although China's approach is full of missteps and inefficiencies, it can and will leapfrog the United States—unless the United States revives the innovation drivers that won the Second World War and the Cold War.

Artificial intelligence applied to big data sets is the core technology of the Fourth Industrial Revolution. If computation is the engine, data is the fuel. While the control point of the twentieth-century economy was oil, the control point of the twenty-first century is data.

As a senior executive of China's Huawei Technologies explained, "This is what we mean by the control point: We don't want to do everything ourselves. If you are a pharmaceutical company, you won't have to duplicate our investment in AI. You simply rent time on the Cloud, using our AI servers, and obtain access to our data. The key is gathering and porting the data to servers where it can be put into usable form. That's our contribution. We don't want to control everything. We want partners who are best in class in every field."[5]

The National Security Commission on Artificial Intelligence wrote in a recent study directed by former Google CEO Eric Schmidt and former Deputy Defense Secretary Robert Work:

> The combination of advances in AI and biology has the potential to reshape the global economy for the next century. Progress in genetic sequencing has given researchers the ability to read the "code of life." Given the significant quantity of data involved, AI will be essential to fully understanding how genetic code interacts with biological processes. Finally, advances in synthetic biology and genetic editing will give researchers the ability to manipulate this code to perform specific functions. Together, these techniques will enable transformational breakthroughs in biology and underpin most future scientific breakthroughs related to human health, agriculture, and climate science. The nation which is best able to simultaneously leverage both technologies will have substantial strategic advantages for the foreseeable future, potentially becoming a global leader in pharmaceuticals, reducing

its reliance on foreign supply chains, and even ensuring it has a healthier and more capable population. These technological breakthroughs will also cause the biotechnology sector to become a major driver of overall U.S. economic competitiveness.[6]

The military applications of the Fourth Industrial Revolution technologies are also transformative. As the National Security Commission on Artificial Intelligence concluded:

> Defending against AI-capable adversaries operating at machine speeds without employing AI is an invitation to disaster. Human operators will not be able to keep up with or defend against AI-enabled cyber or disinformation attacks, drone swarms, or missile attacks without the assistance of AI-enabled machines.[7]

Drone swarms communicating with 5G and guided by AI can do things that manned fighter aircraft cannot do. In addition, swarms of inexpensive sea drones (unmanned underwater vehicles) combined with quantum magnetometers and satellite-based optical sensors may render submarines as vulnerable as were battleships to carrier-based aircraft in the mid-twentieth century.

OUTCOMPETING AMERICA DOMESTICALLY

China has advanced considerably farther than America in implementing and accelerating the fruits of this new revolution. Great advantage will be reaped particularly in transportation and commercial efficiencies.

Mobile broadband is the enabling technology for the Fourth Industrial Revolution, just as railroads were the enabling technology for the First Industrial Revolution.

With the advent of fifth generation (5G) broadband and its capacity to transmit very large amounts of data quickly and with nearly instant response time, the Fourth Industrial Revolution is already in full development. China had installed 800,000 5G base stations as of February 2021, with another 600,000 to 800,000 planned for 2022; these will cover all the major cities in China. The average download speed on China's networks is three hundred ambits per second vs. a US average of fifty-five ambits per second.[8]

The rubric "smart cities" encompasses a complex of 5G-enabled technologies that will drastically reduce freight and package delivery times, passenger waiting time, labor costs, and energy utilization. With high-speed 5G available in almost all China's major urban centers now, China already has introduced autonomous vehicles for urban personal transport. Most Chinese cities are new with main thoroughfares that can easily accommodate autonomous vehicles. The low latency (very rapid response time) of 5G allows autonomous vehicles to communicate with each other almost instantly, a critical safety feature. Following this, China has begun to control urban traffic flow with central computers that match passengers and packages to vehicles, including drones.

Moreover, in terms of commercial efficiency, China has made operational fully automated warehouses, pioneered by the Chinese internet retailers Alibaba and JD.com. It has integrated urban hubs with suburban spokes through high-speed trains. And it has used the Internet of Things (IoT) and "smart" solar panels to reduce the energy cost of heating and cooling buildings.

According to Chinese industry sources, five thousand private industrial 5G networks already are in place in China, with another fifty thousand expected to be completed in the next year.[9] These include the automated ports in Shanghai's Yangshan Container Port, which is the world's largest, as well as industrial robotics, autonomous vehicles, and other applications. Comparable networks in the United States and Europe are, for the most part, experimental rather than operational. The Trump administration's sanctions against sale of components including semiconductors made with US equipment or intellectual property appear to have slowed China's 5G rollout only slightly.

Artificial intelligence allows robots to "learn." CNN reported, for example, in June 2021 as follows:

> In a Hong Kong warehouse, a swarm of autonomous robots works 24/7. They're not just working hard, they're working smart; as they operate, they get better at their job.... The Autonomous Mobile Robots were developed by Chinese startup Geek+. As they move around the warehouse, they're guided by QR codes on the floor, and using AI they are able to make their own decisions, including what direction to travel and what route to take to their destination.[10]

China did not invent industrial robotics, but it leads the world in robot usage with 140,000 in place in 2020. That is more than Japan, the United States, South Korea, and Germany combined. With the introduction of 5G-area as well as private networks, China is poised for a productivity quantum-leap in manufacturing, transportation, and logistics.

As the National Security Commission on Artificial Intelligence has concluded:

> The race to research, develop, and deploy AI and associated technologies is intensifying the technology competition that underpins a wider strategic competition. China is organized, resourced, and determined to win this contest. The United States retains advantages in critical areas, but current trends are concerning.[11]

Technological innovation harnessed to production capacity produces a snowball effect. As innovations are commercialized, manufacturing and logistics industries provide a real-world laboratory to test and deploy new innovations. America invented the digital age because most of the decisive innovations—optical networks, inexpensive semiconductor manufacturing, displays, sensors, and communications technology including the internet—emerged from corporate laboratories linked to production capacity. China already produces about 30 percent of the world's manufactured products and is moving up the technology spectrum. It also produces seven times as many STEM baccalaureate degrees as the United States. China is close to achieving a critical mass of skills, technology, and supply-chain depth that will leave the United States behind—just as the United States left Britain behind during the twentieth century.

SINO-FORMING THE WORLD

One only need look to Latin America, America's geopolitical backyard, for indications of how fast Chinese influence has grown. In 2019 Mexico had seventy-seven mobile broadband accounts per one hundred people, whereas in 2013 it only had twenty-three such accounts.[12] And Mexico last year had the world's highest percentage growth in e-commerce. In 2020, about a third of Mexico City drivers were using the Waze navigation app, making Mexico its third largest market, compared to virtually zero in 2015. Huawei (with some participation by Nokia) built Mexico's mobile broadband infrastructure.

In Brazil, regulators have approved Huawei's participation in the auction of 5G spectrum, making the Chinese telecom giant the probable dominant player in the largest South American economy. Apart from the public broadband network, Huawei is building dedicated 5G networks for Brazilian soybean farms that link sensors at the individual plant level to drone deliveries of water, fertilizer, and pesticide.

There is nothing inherently wrong in Chinese investments that raise productivity and living standards in the developing world. What the United States should fear most is not what China does wrong, but what it does right. For example, the United States in 2020 exported $25.7 billion worth of soybeans, America's single largest export item for which China is the largest buyer. "Smart" agriculture in Brazil and other large soy producers will rapidly reduce China's dependency on imports from the United States and erode American exports at a moment when the US current account deficit is running at about $1 trillion a year.

In a much broader sense, the scarcest resource in the world economy during the next several decades will be working-age adults. Some Western analysts point to China's shrinking demographics as a harbinger of

China's future decline, but China's demographic problems are hardly unique. Its projected elderly dependent ratio (the number of retirees per person of working age) will remain at about the same level as Germany's and well below that of Japan and South Korea. The United States, with a total fertility rate of just 1.7, will have a dependent ratio only slightly lower than China's.

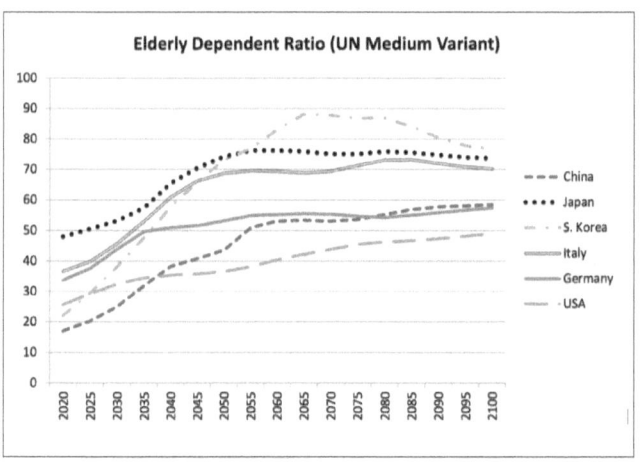

There are several ways to mitigate the economic effect of an aging population. The two most important are (1) to increase labor productivity, and (2) to export capital to countries with a large supply of labor. Contrary to China, the United States and much of Western Europe have pursued a different strategy: importing and giving citizenship to an ethnically dissimilar pool of laborers.

As noted, China's advances in labor saving will make it possible to eliminate many labor-intensive jobs. The job description that encompasses the most workers in the United States is "driver." China's newly constructed urban infrastructure and 5G buildout supports autonomous vehicles far better than the aging, often chaotic infrastructure of American cities. China, moreover, is ahead of the United States in warehouse automation. E-commerce accounts for more than 50 percent of all retail sales in China, compared to 14 percent in the United States as of the first quarter of 2021.[13]

Even more important are China's inroads into the developing world. China's $2 trillion Belt and Road Initiative (BRI), in combination with digital technology, aims to integrate billions of people in the developing world into China's economic sphere. Cheap telecommunications make it possible to deliver financial services and world market access to people who previously were cut off from the global economy.

As I have written in *You Will be Assimilated*:

> The backward economies in which most of the world's people live waste time and grind down the spirit. In so-called emerging markets, between one-third and two-thirds of the workforce spend most of their time doing little or nothing. The proportion of workers outside the formal economy ranges from 32 percent in Turkey to nearly 60 percent in Nigeria.
>
> Everyone has a hustle, but no one has access to a capital, nor rights to property, nor security from the arbitrary intrusion of the authorities. Men work seasonal construction jobs for cash payments, and women find things to sell. No one pays taxes; there is no way to collect them, and "informal" workers couldn't afford them in any case. Governments in turn provide paltry services and accumulate debts. An inbred elite milks

the state budget and manages monopolies. China is the first emerging country to fully integrate the informal economy into the formal economy, and it has done so through smartphones and electronic payments. China is on the way to becoming a "cashless society," in which virtually all transactions are electronic and therefore transparent to the tax authorities. Turkey, which has the same rate of smartphone penetration as China, plans to become a cashless society by 2023.[14]

BROAD ECONOMIC CONSEQUENCES FOR AMERICA

Digital technology is a winner-take-all world. There is room for one Microsoft, rather than Excel along with Lotus 123; one Google, rather than Google along with Altavista or Bing; one Facebook, rather than Facebook along with Myspace; one Amazon, and so forth.

If China creates the dominant firms with AI/big data applications, trillions of dollars of wealth will migrate to China from the United States. The capacity of American firms to maintain high levels of R&D and preserve America's technological edge will erode rapidly.

The economic future of several billion people outside the economic mainstream is also at stake. At its June 2021 meeting in England, the Group of Seven offered an unconvincing response to the challenge from China's

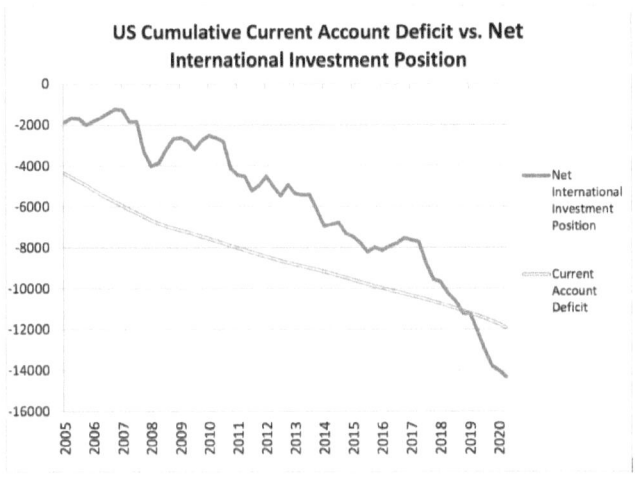

Belt and Road Initiative by reiterating a preexisting commitment to provide $100 billion of funding for climate-related investments in the developing world. That is a small amount compared to the $2 trillion that China proposes to spend on its initiative, and it is focused on an area of investment that is not the most urgent priority for most developing countries.

Moreover, America's capacity to sustain a federal debt of $24 trillion (not to mention unfunded Social Security and Medicare liabilities of perhaps $100 trillion) will erode, along with the value of the US dollar.

Between 1998 and 2021, the United States ran a cumulative current account deficit of about $11 trillion, and its net international investment position fell to minus $13.3 trillion. Foreigners, that is, invested in the United States roughly the same amount that the United States owed to foreigners for imports of goods and services in excess of exports.

The United States is able to run budget deficits now approaching 15 percent of gross domestic product, a level without precedent in peacetime, in part because of what economists call "seigniorage," which is named after the premium that a monarch earned by coining precious metal into currency. The United States accounts for about 8 percent of world exports, but more than 50 percent of international reserves and offshore-backed deposits.[15] China already is the world's biggest exporter with 12 percent of the total, and it will become the world's largest economy in dollar terms before the end of this decade. If China's currency gains a global role commensurate with its economic standing, the dollar's reserve role will fade like the pound sterling before it, and the American capacity to borrow overseas will shrink considerably. That, in turn, implies a severe adjustment for a heavily geared US economy.

China's development of a "digital yuan" has prompted speculation about the Chinese currency as a possible replacement for the US dollar. But China has neither the capacity nor the intention to replace the dollar system as it now stands. Rather, it is building another system in parallel that is likely to erode the existing dollar system and reduce America's capacity to finance its deficit.

This will take place through digital currencies, which promise to drastically reduce transaction costs for international trade financing while improving transaction security. In the eighteenth century, financier Nathan Rothschild apocryphally said a bill of exchange in international trade should taste of salt, after accompanying cargo on a sea voyage. Blockchain allows the tracking of goods from factory to warehouse to port to container to ship, and it enables just-in-time deliveries along with just-in-time payments. Once international payment mechanisms are in place, China can prevail on its trading counterparties to conduct transactions in the digital yuan.

The United States and other Western countries can offer their own digital currencies, to be sure, but China's advantage lies not on the monetary medium as such but rather in the transformation of trade and logistics brought about by the Fourth Industrial Revolution. If China dominates "smart logistics," the digital yuan will emerge as the leading unit of account in world trade.

World trade financing today requires banks to accept bills of exchange in trade, and importers maintain tens of trillions of dollars in bank balances as collateral for their overseas orders. The combination of "smart logistics," in which all commodities in trade carry a digital signature at every stage of production and transportation, and digital payments, will reduce working capital in international trade substantially. That is not a small matter. During the sixteenth century an enormous amount of the Spanish Empire's working capital was tied up in the mining, storage, shipment, and security of gold and silver bullion, which Spain extracted from the New World and (mostly) used to pay for luxury imports from Asia. That was a primitive precious metals standard in which the exchange of actual bullion financed the Spanish Empire. The central banking system developed by the Netherlands and Britain, in which the central bank undertook to maintain the parity of the currency against bullion while conducting most trade through bills of exchange "accepted" by banks, represented an enormous increase in efficiency. The next generation of trade finance in the Fourth Industrial Revolution will bring about yet another jump in efficiency.

Efficiency in trade finance means simply that less working capital will be required to support transactions. That is a good thing for most factors of production, but it is not a good thing for the United States under present conditions, because a great deal of the working capital now dedicated to international trade constitutes an interest-free loan to the United States. The US$16 trillion deposited in banks as working capital for international trade is invested in US money markets or US government securities at an effective interest rate of close to zero, thanks to the Federal Reserve's expansionary monetary policy. Directly or indirectly, these deposits support the market for US government debt.

Smart logistics and digital payments systems are already transforming the role of banking. As J. P. Morgan Chase CEO Jaime Dimon wrote in his 2021 letter to shareholders,

> It is completely clear that, increasingly, many banking products, such as payments and certain forms of deposits among others, are moving out of the banking system. In addition, lending in many forms—including mortgage, student, leveraged, consumer and non-credit card consumer—is moving out of the banking system. Neobanks and nonbanks are gaining share in consumer accounts, which effectively hold cash-like deposits. Payments are also moving out of the banking system, in merchant processing and in debit or alternative payment systems.[16]

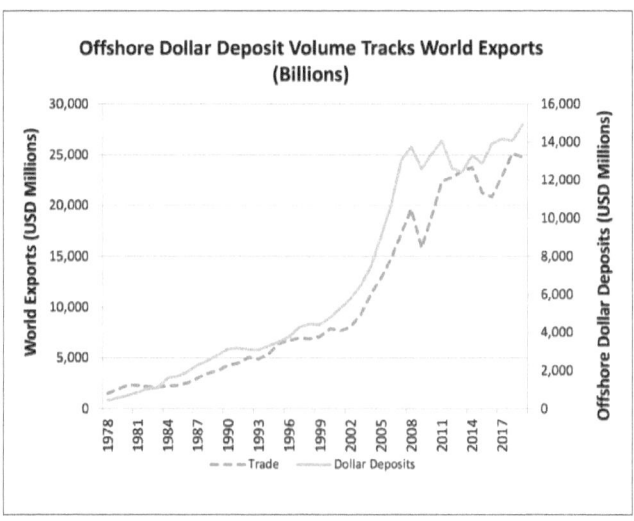

The chart shows that the size of cross-border bank deposits (as reported by the Bank for International Settlements) tracked the volume of world exports during the past forty years, rising rapidly during the early 2000s as trade growth accelerated and leveling off after the 2008 financial crisis as trade stagnated.

The Bank for International Settlements breaks down the volume of cross-order deposits by currency, and the dollar dominates, with more than $16 trillion outstanding.

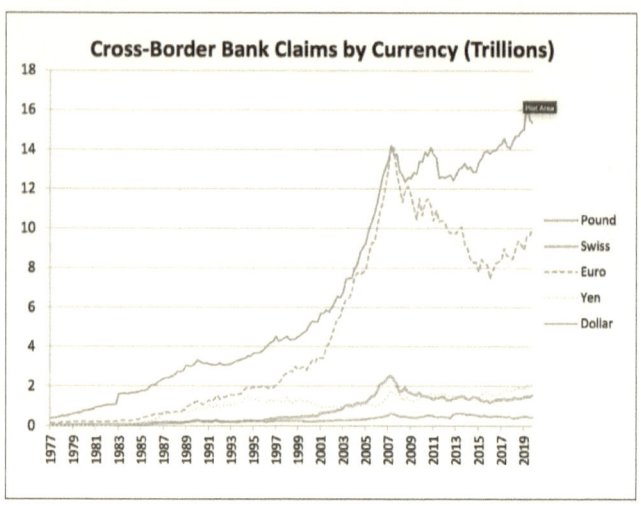

These deposits, mostly the collateral for international transactions in goods, services, and securities, amount to a $16 trillion low-interest or interest-free loan to the United States. For example, an importer in Wisconsin who wants to buy widgets from Wuhan, China establishes a line of credit with a bank. The widget manufacturer's bank verifies that the importer in Wisconsin has funds to back up an order, and it issues a credit line to the manufacturer. After a number of months the widgets are manufactured, shipped, and delivered to the importer, at which point the importer's bank remits funds to the manufacturer's bank. The cost of the widgets is tied up in working capital for months. Smart logistics and the Internet of Things can verify every purchase of raw materials, every movement to a warehouse, and then to a shipping container, a port, and a truck en route to final delivery. Production, warehousing, and shipment become completely transparent, so that lenders can finance each stage of the process without requiring large bank deposits as security. The amount of working capital in the system will shrink drastically.

In addition, foreigners own about $8 trillion of US treasury securities, most of which are held by foreign central banks as currency reserves. Together with the transaction balances in the banking system, total overseas dollar holdings amount to more than $22 trillion, or more than a full year of the United States' gross domestic product.

America's dependence on the foreign credit made possible by the dollar's reserve role has risen sharply relative to the size of its economy, from about 20 percent of GDP in 1978 to 110 percent of GDP today. Countries hold reserves against the eventuality that an economic shock of some kind may require the central bank to lend funds that can be used to make international payments. At present, about four-fifths of international trade is conducted in US dollars, and about three-fifths of central bank reserves are held in US dollars.[17] The advent of smart logistics and digital payments systems will shift transactions in world trade to the currencies of countries that do the most trade, starting with China and its digital yuan. Central banks will hold fewer reserves in dollars.

A trend toward a reduced use of the dollar is already evident. As the World Economic Forum wrote in May 2021:

> The share of US dollar reserves held by central banks fell to 59 percent—its lowest level in 25 years—during the fourth quarter of 2020, according to the IMF's Currency Composition of Official Foreign Exchange Reserves (COFER) survey. Some analysts say this

partly reflects the declining role of the US dollar in the global economy, in the face of competition from other currencies used by central banks for international transactions. If the shifts in central bank reserves are large enough, they can affect currency and bond markets.[18]

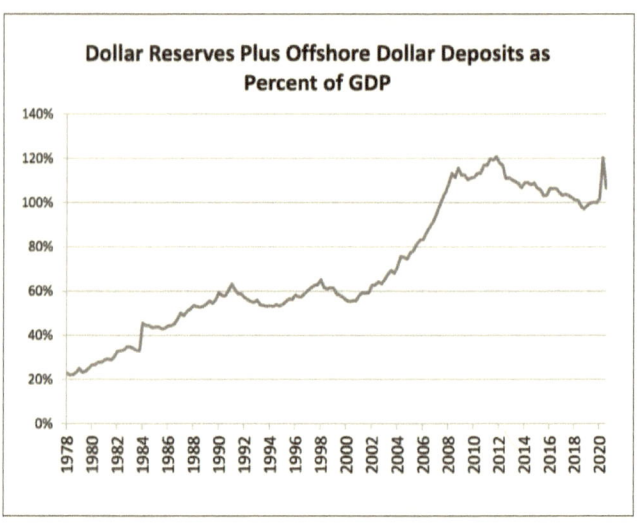

Overseas bank deposits denominated in US dollars are typically invested in liquid US securities, including US treasuries, corporate bonds, and federal agency bonds (mainly mortgage-backed securities). Banks do not typically buy stocks for their portfolios because risk capital requirements are prohibitive. The chart below shows the cumulative foreign purchases of US securities by foreigners since 1978 as reported by the US Treasury. The total of about $13 trillion is roughly equal to America's net foreign investment position.

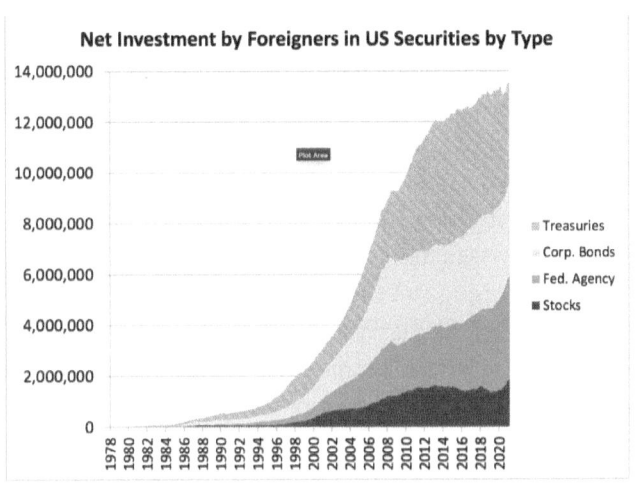

A substantial reduction in offshore dollar deposits owing to structural changes in the financing of world trade would cause a corresponding reduction in foreign holdings of US securities. Given the size of the US federal deficit, such a reduction in demand for US securities could be painful. The United States stock market now trades at nearly thirty times earnings, a multiple not seen since 2000, before a long and painful correction. The lofty valuation of the US equity market is driven by the longest period of negative real interest rates in US history. If the dollar's reserve status is compromised, the United States will no longer be able to borrow at negative real rates, and rising bond yields will put pressure on equity markets, depressing the value of the US stock market and reducing the value of pension and retirement funds.

At current projections, the US Treasury now has to finance a federal deficit roughly equal to 10 percent of US GDP. During the past year, the Federal Reserve and

US commercial banks have financed virtually all the issuance of US Treasury securities. The decline of the dollar's reserve-currency role means that foreign central banks would stop investing in US Treasury securities and liquidate a considerable portion of the Treasury securities they presently own. That would force the Federal Reserve to buy even more Treasury securities ("monetize the debt" by creating currency with which to buy these securities), leading to aggravated inflation, or, as an alternative, to raise interest rates high enough to attract the world's capital to US Treasury securities.

What does national decline look like? Per capita GDP in the United States and the United Kingdom were roughly equal in 1930, despite Britain's considerable population loss during the First World War. Today US per capita GDP is 45 percent higher than the UK's, according to data compiled by the Maddison Historical Statistics Project. The loss of Britain's industrial

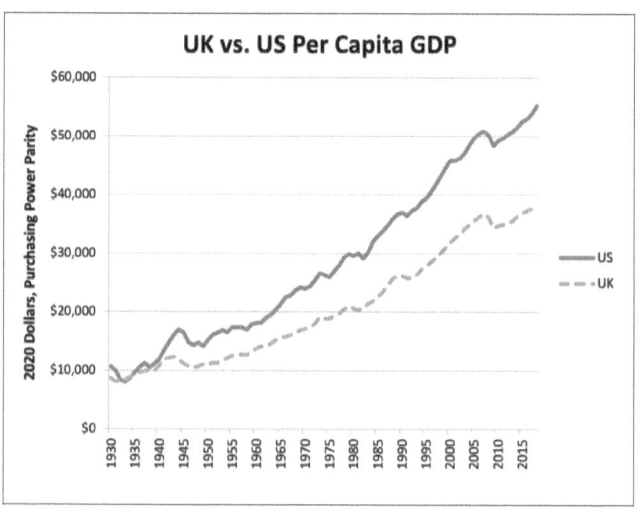

preeminence, its empire, and its global reserve currency did not ruin Britain, but it left Britons far less wealthy than Americans. Britain, moreover, found a niche in the Atlantic Alliance, and became the leading provider of financial and related services within the dollar sphere. The British, as Harold Macmillan supposedly said, "are Greeks in this American empire."[19]

Americans no doubt will find remunerative things to do in a new Chinese Empire. The Chinese do not conquer and destroy. They assimilate. They are incurious about how barbarians manage their internal affairs, contemptuous of democracies who do not elevate their cleverest exam-scorers to Mandarin positions. America will persist even if it doesn't prevail. We will still write smartphone aps. We will be the Greeks in a new Roman Empire.

WHAT SHOULD THE UNITED STATES DO?

The United States has faced technological challenges in the past. The Eisenhower and Kennedy administrations responded to Russia's launch of Sputnik in 1957 and Yuri Gagarin's first manned space flight in 1961 with the promise to land a spaceship on the moon by the end of the 1960s. The Carter and Reagan administrations responded to Russia's military buildup in Europe and Russian advances in anti-aircraft rocketry with decisive advances in military technologies and the promise of a missile shield under the Strategic Defense Initiative. The United States achieved a degree of technological superiority unprecedented in the history of modern warfare.

America also invented the digital age. Defense Department funding, mainly through the great corporate laboratories at AT&T, General Electric, RCA, IBM, and others led to the internet, CMOS mass-manufacturing of powerful computer chips, plasma, and LED displays, the semiconductor laser and its application to optical networks, the Graphic User Interface, and countless other game-changing technologies. It is no exaggeration to say that all the key inventions that together comprised the Digital Revolution began at the Defense Advanced Research Projects Agency.

During the Reagan years, federal R&D funding rose to 1.4 percent of GDP. It now stands at roughly half that level. We need to regain that level of support for basic research, but we have to do it in the right way. Both political parties want to sustain US competitiveness by means of funding university research laboratories through the National Science Foundation. That is not enough. If academic scientists publish their research in academic journals but entrepreneurs fail to turn this research into new products, the benefit of the research will accrue to whoever reads the academic journals—for example, China.

For the past twenty years, US corporations have shifted away from capital-intensive manufacturing industries toward "capital light" software and services. At the same time, our Asian competitors have increased the capital intensity of their economies. The capital intensity (the ratio of total assets to earnings before interest and taxes) of the S&P 500 Index stock index has changed little. During the same period, the capital intensity of the components of China's Shanghai Composite Index has nearly tripled. South Korea's KOSPI stock index shows a capital intensity roughly equal to China's.

Asian industry is more capital-intensive because Asian governments subsidize capital-intensive industry. Taiwan Semiconductor Manufacturing Corporation became the world's dominant fabricator of computer chips because Morris Chang, then the chief engineer at Texas Instruments, persuaded the Taiwanese government to subsidize his startup in 1987. The United States offers some tax subsidies for R&D, but American subsidies are less than half the international level. The US taxes capital income at an all-in rate of 34 percent, according to the Tax Foundation, compared to a rate of 27 percent for South Korea.[20] The tax system should support capital-intensive investment and R&D. The Trump administration's corporate tax cut in 2018 reduced the headline corporate tax rate, a measure much favored by corporate lobbyists and Republican think tanks, but its economic benefit was modest. US business used the additional cash flow from the tax cut to buy back stocks, so that equity buybacks in 2019 exceeded capital investment for the first time since 2008.[21]

Industrial subsidies typically foster cronyism, corruption, and inefficiency. The United States fielded a successful industrial policy in the past for two reasons. First, the federal government (above all the Department of Defense) set clear priorities dictated by military necessity. Superior weapons systems or predominance in space demand real breakthroughs at the frontiers of science. Second, we drew a bright line between the responsibilities of the public sector and those of the private sector. The federal government paid for basic research by corporate laboratories, but the private sector shouldered the risk of commercialization. The United States should not directly invest in private industries except in the special case of products whose strategic importance is beyond question—for example, in semiconductors. The right combination of R&D subsidies and tax incentives should be sufficient to persuade American corporations to revive the system of corporate laboratories that collaborated so well with the Defense Department during the 1970s and 1980s.

Moreover, offshoring of US manufacturing facilities was motivated in large measure by the lower cost of overseas labor. Technological changes in manufacturing now favor the return of manufacturing capacity to the United States. As venture capitalist Henry Kressel wrote in 2020,

> Manufacturing has been out of fashion in the US as much of it migrated offshore in search for lower costs. But technology, particularly artificial intelligence, is changing that calculation for high-tech industries that can use extensive robotics. This change makes manufacturing closer to home practical for some industries and creates a competitive advantage by allowing a much more nimble business compared with relying on offshore production.[22]

The United States still can lead the Fourth Industrial Revolution. But we do not have a lot of time to lose. China is close to attaining a critical mass of talent, skills, technological capacity, and logistical depth with a population nearly five times that of the United States. At some point in the foreseeable future, the United States will not be able to catch up.

Despite its best intentions, the Trump administration's economic initiatives—the corporate tax cut and the tariffs against Chinese imports—did little to revive US manufacturing. Industrial production shrank in 2019—before the COVID-19 recession. The Biden administration's massive fiscal stimulus meanwhile has given us a burst of inflation that squeezes manufacturers' margins and deters investment. The Bureau of Labor Statistics reported the following on July 14, 2021: "For the 12 months ended in June, prices for processed goods for intermediate demand jumped 22.6 percent, the largest increase since rising 23.6 percent for the 12 months ended in February 1975."[23] What we require is a radical return to the successful policies of the past: A defense driver for technological discovery, public-private partnerships for R&D, and a tax structure that favors manufacturing.

ABOUT THE AUTHOR

David P. Goldman is a Washington Fellow at the Claremont Institute's Center for the American Way of Life, as well as the president of Macrostrategy LLC. He writes the "Spengler" column for Asia Times Online and the "Spengler" blog at PJ Media, and is the author of several books, including, *You Will Be Assimilated: China's Plan to Sino-Form the World* (Bombardier Books) and *How Civilizations Die (and Why Islam is Dying Too)* (Regnery).

ENDNOTES

1. Klaus Schwab, "The Fourth Industrial Revolution: What It Means, How to Respond," World Economic Forum, January 14, 2016, https://www.weforum.org/agenda/2016/01/the-fourth-industrial-revolution-what-it-means-and-how-to-respond/.
2. See, for example, Eric Schmidt and Robert Work, Final Report of the National Security Commission on Artificial Intelligence, March 2021, https://www.nscai.gov/wp-content/uploads/2021/03/Full-Report-Digital-1.pdf; Peter Cowhey, Meeting the China Challenge: A New American Strategy for Technology Competition, The 21st Century China Center of the University of California San Diego/The Asia Society Center on US-China Relations, November 2020, https://asiasociety.org/sites/default/files/inline-files/report_meeting-the-china-challenge_2020.pdf.
3. Over the past thirty-five years the United States sustained a cumulative current account deficit of $12.5 trillion, matched by a net foreign asset position of negative $13 trillion.
4. Edward Yardeni and Joe Abbott, "Stock Market Briefing: FAANGMs," Yardeni Research, Inc., August 18, 2021, https://www.yardeni.com/pub/faangms.pdf.
5. David P. Goldman, *You Will Be Assimilated: China's Plan to Sino-Form the World* (New York: Bombardier, 2020), 77.
6. Eric Schmidt and Robert Work, "Final Report," National Security Commission on Artificial Intelligence, 584, accessed August 19, 2021, https://www.nscai.gov/wp-content/uploads/2021/03/Full-Report-Digital-1.pdf.
7. Ibid., 9.
8. Mike Dano, "A Direct Comparison: Tabulating 5G in the USA and China," Light Reading, December 16, 2020,

9. https://www.lightreading.com/security/a-direct-comparison-tabulating-5g-in-usa-and-china/d/d-id/766196.
10. David P. Goldman, "US is Chasing China's Tail on 5G," Asia Times, May 30, 2021, https://asiatimes.com/2021/05/us-is-chasing-chinas-tail-on-5g/.
11. Stephanie Bailey, "This Swarm of Robots Gets Smarter the More It Works," CNN, June 16, 2021, https://edition.cnn.com/2021/06/15/asia/swarm-robots-hong-kong-warehouse-hnk-spc-intl/index.html.
12. "Final Report," National Security Commission on Artificial Intelligence, 11.
13. David P. Goldman, "Sino-forming South of the US border," Asia Times, February 12, 2021, https://asiatimes.com/2021/02/sino-forming-south-of-the-border/.
14. "Ecommerce Share of Retail Sales (2019–2024)," Oberlo, accessed August 19, 2021, https://www.oberlo.com/statistics/ecommerce-share-of-retail-sales; Michael Keenan, "Global Ecommerce Explained: Stats and Trends to Watch in 2021," Shopify.com, May 13, 2021, https://www.shopify.com/enterprise/global-ecommerce-statistics.
15. Goldman, *You Will be Assimilated*, 68–69.
16. The current share of US exports on the world market (8 percent) represents a decline from 14 percent in 2000 and 11 percent in 2015. This data is from "Percent of World Exports—Country Rankings," theGlobalEconomy.com, accessed August 19, 2021, https://www.theglobaleconomy.com/rankings/share_world_exports/.
17. Jamie Dimon, "Chairman & CEO Letter to Shareholders," JPMorgan Chase & Co., Annual Report 2020, April 7, 2021, https://reports.jpmorganchase.com/investor-relations/2020/ar-ceo-letters.htm#banks-enormous.
18. See, for instance, "Currency Composition of Official Foreign Exchange Reserves," International Monetary Fund, accessed August 19, 2021, https://data.imf.org/?sk=E6A5F467-C14B-4AA8-9F6D-5A09EC4E62A4.
19. Serkan Arslanalp, "Are We Witnessing the Decline of the Dollar's Global Significance? Here's What the Foreign Exchange Reserves Tell Us," World Economic Forum, May 12, 2021,

19. https://www.weforum.org/agenda/2021/05/us-dollar-share-of-global-foreign-exchange-reserves-drops-to-25-year-low/.
19. Quoted in Nigel John Ashton, Kennedy, Macmillan and the Cold War: The Irony of Interdependence (New York: Springer, 2002), 6.
20. Jacob Lundberg and Johannes Nathell, "Taxing Capital—An International Comparison," Tax Foundation, May 11, 2021, https://taxfoundation.org/tax-burden-on-capital-income/.
21. Lu Wang, "Stock Buybaks Top Capex for First Time Since 2008," Bloomberg News, March 3, 2019, https://www.bloomberg.com/news/articles/2019-03-03/stock-buybacks-top-capex-for-first-time-since-2008-citi-says.
22. Henry Kressel, "Flexible High-Tech Manufacturing Is the Future," Asia Times, May 12, 2020, https://asiatimes.com/2020/05/flexible-high-tech-manufacturing-is-the-future/.
23. U.S. Department of Labor, U.S. Bureau of Labor Statistics, PPI Detailed Report: Data for June 2021 26, no. 6, ed. Joseph Kowal, Serah Hyde, Gabriel Vera, and Timothy Schermerhorn, accessed August 19, 2021, https://fraser.stlouisfed.org/files/docs/publications/ppi/ppi_bls_202106.pdf

www.ingramcontent.com/pod-product-compliance
Lightning Source LLC
Chambersburg PA
CBHW031929240526
45464CB00023B/2881